May you always
have the heart of a
Superhero and SOAR!

♥ Be Strong!

Emily

SOARING
ADVENTURES OF AN AERIAL APPLICATOR

WRITTEN BY

Bethany Morton-Gannaway

ILLUSTRATED BY

Vera K. Linton

ISBN Number: 978-0-9882112-5-4

Published in the United States by Whimsy House Publishing
4763 W. Spruce Avenue, Suite 109
Fresno, CA 93722

Whimsy House Publishing

a division of:
The Magical Mailbox

www.playcreategrow.com

DEDICATION

For my Dad.

Thank you for teaching me:

That life is always getting bigger, better, and brighter.

To always go with your first instinct.

To dream big.

And that God gets all the glory.

And a special thanks to Vera-
Thank you for capturing my dream. Your art has become a voice for the aerial application industry. Together I know we will be loud! Love you, sweet friend.

It wouldn't be strength, invisibility, or doing things right on the first try.

Only one I would want. The power to fly.

I want to be a superhero who helps people too.

Maybe fight fires or give hungry people food.

I could give clothes to people who need to stay warm.

Or I could fight off a scary insect swarm.

I could clean up the ocean from spills that hurt the coral and fish.

I want to do it all! That is my wish!

"Excuse me," said a bright yellow airplane.

He looked different than a jet, but a lot was the same.

I know a way you can do all that and more!

Be a pilot and in the sky you will soar.

You can be a superhero who helps all of the earth.

Showing everyone what the world is worth."

"How, Mr. Airplane? What must I do?"

"Jump in my cab and I will show everything to you!"

I stared at the gauges and the propeller up ahead.

Listening to all that Yellow Airplane said.

"Aerial Applicators are some of the greatest heroes around.

They help people every time they lift off from the ground!

When there is a wildfire burning trees big and tall,

Aerial Applicators come and answer the call.

When food crops need help growing so that people can eat,

Aerial Applicators fly on fertilizer-nothing is too big of a feat!

Cotton clothing takes more than a needle and thread.

Aerial Applicators prepare cotton for blankets warm on your bed.

Like heroes they guard fields from insects eating our crops.

Aerial Applicators safely make the swarms come to a stop.

When the oceans are polluted by black oil and harmful things,

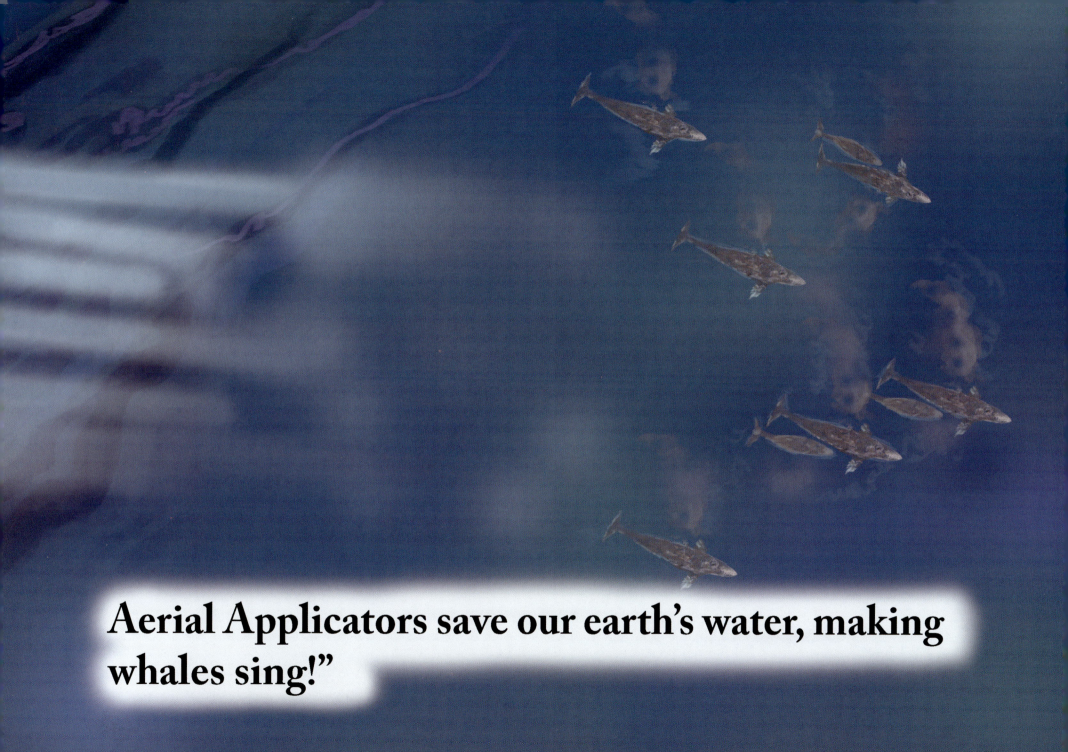

Aerial Applicators save our earth's water, making whales sing!"

As I looked down below, it was a beautiful sight.

Crops of almonds and lettuce, tomatoes red and bright.

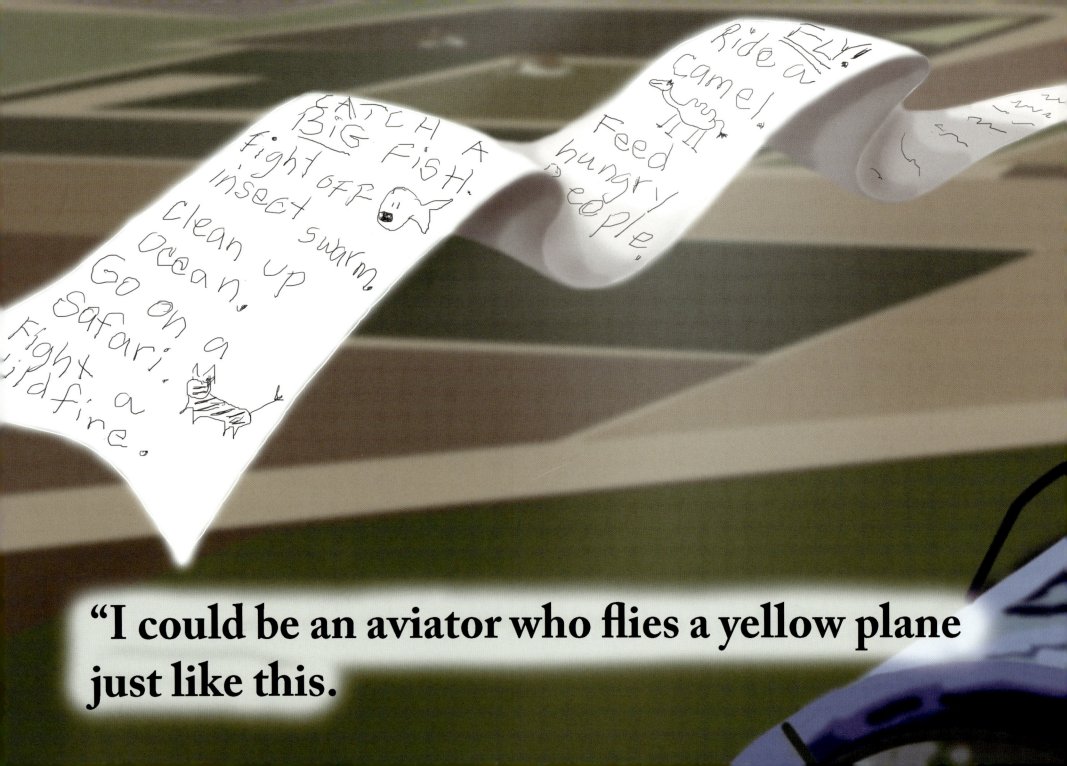

"I could be an aviator who flies a yellow plane just like this.

And still check all the dreams off my list!"

Yellow Airplane began to line up and come in to land.

"If you want to do this I know you'll be grand!

Aerial Applicators must go to school and get a pilot's license too.

When you grow up I will be waiting here for you."

As the years passed I got older and grew,

Learning about planes and agriculture-things old and new.

The most important thing I learned when my flying days came,

Is that Aerial Applicators and superheroes are one and the same.

This book belongs to: